THIS BOOK BELONGS TO:

ISBN: 979-8-9865130-5-8

THE GREAT MATH DETECTIVE'S WORKBOOK

HI FRIENDS!
WE ARE A TEAM OF MATH DETECTIVES
AND WE LOVE SOLVING MATH MYSTERIES.
THE COOLEST PART IS THAT THE
MYSTERIES WE SOLVE ARE NOT ONLY
IN THIS WORKBOOK. BUT ALL AROUND US!

AS A GREAT MATH DETECTIVE,
OBSERVATION AND SKILL-BUILDING
ARE WHAT MAKE OUR TEAM STRONG.
SO, WE WILL ASK YOU TO PRACTICE
WITH US. BEFORE WE GET STARTED,
TELL US YOUR NAME BY WRITING IT
ON THE NEXT PAGE. AND ALWAYS
REMEMBER IF YOU NEED HELP,
IT'S OKAY TO ASK A GROWN-UP.

HELLO! WHAT'S YOUR NAME?

Write your name below with a...

PENCIL

CRAYON

PEN

MARKER

HELLO! WHAT'S YOUR PHONE NUMBER?

Write your phone number with a...

PENCIL

CRAYON

PEN

MARKER

ALL ABOUT ME

Make a self portrait below using different art materials.

THE STORY OF YOU

In the boxes below, share your story! What important events have happened in your life? What are your strongest memories? What hobbies or passions are important to you?

Where were you born?

My favourite thing to do is...

I find it hard to....

I hope that....

My biggest struggle has been....

My biggest success has been....

When I was little....

LEARNING NUMBERS

1 2 3

4 5 6

7 8 9

10

KEEP GOING, YOU'RE DOING AMAZING!

GREAT WORK!

LEARNING NUMBERS

Trace the numbers below.
And write it on your own on the next line.

1 1 1 1 1 1 1 1 1 1 1 1 1 1 1

1

2 2 2 2 2 2 2 2 2 2 2 2 2 2 2

2

3 3 3 3 3 3 3 3 3 3 3 3 3 3 3

3

4 4 4 4 4 4 4 4 4 4 4 4 4 4 4

4

5 5 5 5 5 5 5 5 5 5 5 5 5 5 5

5

LEARNING NUMBERS

Trace the numbers below.
And write it on your own on the next line.

6 6 6 6 6 6 6 6 6 6 6 6 6 6

6

7 7 7 7 7 7 7 7 7 7 7 7 7

7

8 8 8 8 8 8 8 8 8 8 8 8 8

8

9 9 9 9 9 9 9 9 9 9 9 9 9 9

9

10 10 10 10 10 10 10 10 10 10 10 10 10 10 10

10

LEARNING NUMBERS

Trace the numbers below.
And write it on your own on the next line.

11 11 11 11 11 11 11 11 11 11 11 11 11 11

11

12 12 12 12 12 12 12 12 12 12 12 12 12 12

12

13 13 13 13 13 13 13 13 13 13 13 13 13 13

13

14 14 14 14 14 14 14 14 14 14 14 14 14 14

14

15 15 15 15 15 15 15 15 15 15 15 15 15 15

15

LEARNING NUMBERS

Trace the numbers below.
And write it on your own on the next line.

16 16 16 16 16 16 16 16 16 16 16 16 16 16

16

17 17 17 17 17 17 17 17 17 17 17 17 17 17

17

18 18 18 18 18 18 18 18 18 18 18 18 18 18

18

19 19 19 19 19 19 19 19 19 19 19 19 19 19

19

20 20 20 20 20 20 20 20 20 20 20

20

ADD THE PICTURES

Count, add, and write the correct number in the empty box.

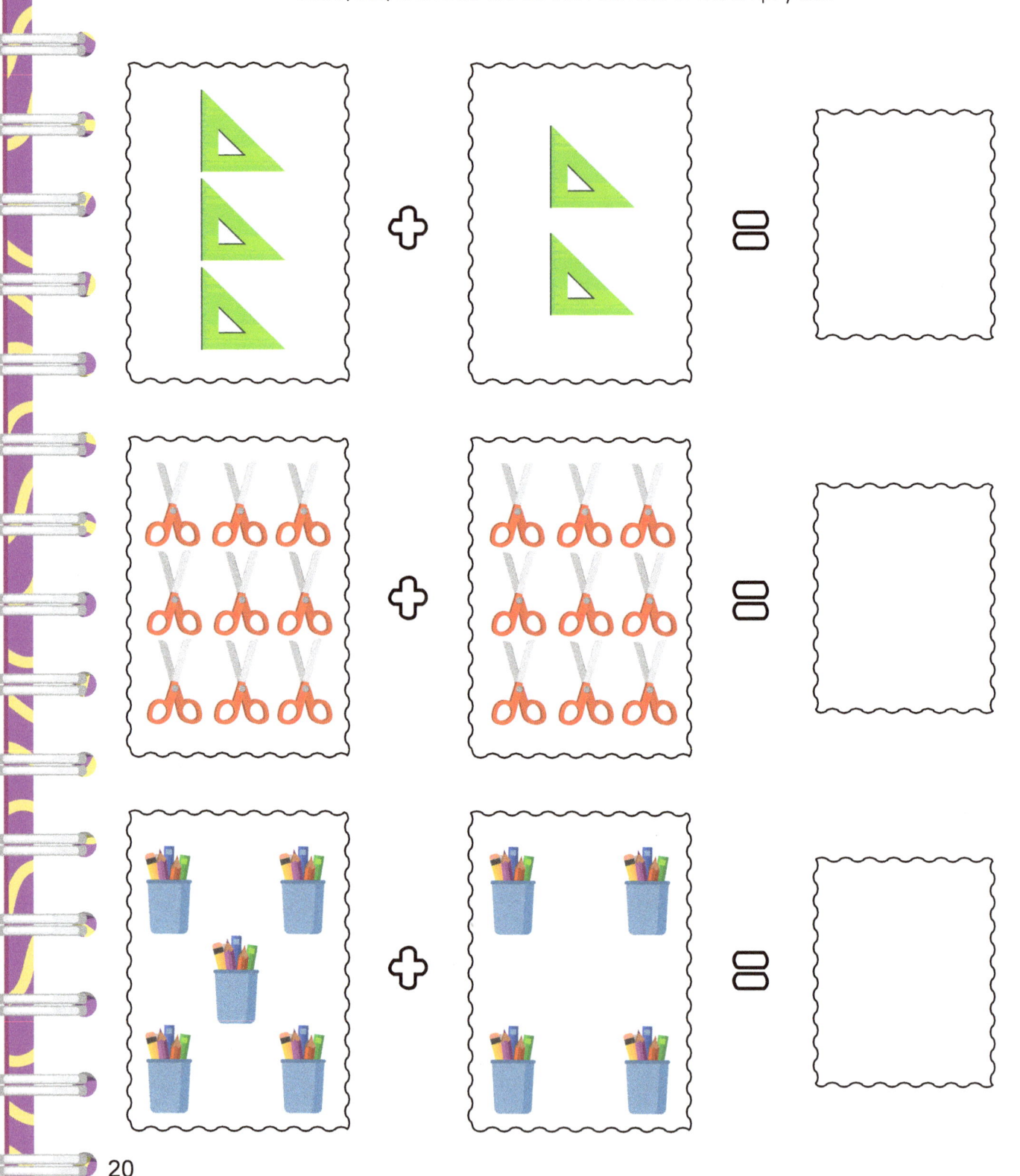

ADD THE PICTURES

Count, add, and write the correct number in the empty box.

MORE AND LESS

In each row, count the number of items and color the image which has more number of objects.

SAME AND DIFFERENT

Circle the items that are the same. Place a square around the items that are different.

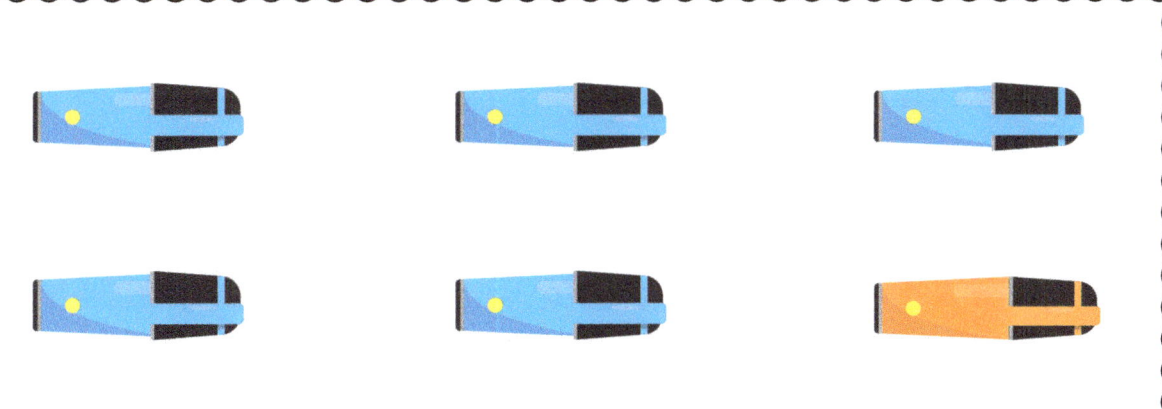

COUNT THE ITEMS

Count each items and circle the correct number.

3
6
5

7
12
20

3
4
9

PHEW! YOU HAVE BEEN WORKING VERY HARD. LET'S STOP FOR A MOMENT TO TAKE A MENTAL BREAK AND STRETCH.

TOUCH YOUR TOES

BREATHE IN YOUR NOSE AND OUT YOUR MOUTH

AND NOW REPEAT!
REACH UP TO THE SKY

TOUCH YOUR TOES

BREATHE IN YOUR NOSE AND OUT YOUR MOUTH

REACH UP TO THE SKY

As a Great Math Detective. It's important to know that sometimes numbers can be written with letters. Here, let me show you:

1 is the same as **one** & **2** is the same as **two**

See? You have what it takes! We're not done yet, though. There are more numbers for us to practice below.

1 one one one one one one

one

2 two two two two two two

two

3 three three three three three

three

4 four four four four four four

four

5 five five five five five five

five

As a Great Math Detective. It's important to know that sometimes numbers can be written with letters. Here, let me show you:

1 is the same as **one** & **2** is the same as **two**

See? You have what it takes! We're not done yet, though. There are more numbers for us to practice below.

6 six   six   six   six   six   six
six

7 seven   seven   seven   seven   seven
seven

8 eight   eight   eight   eight   eight
eight

9 nine   nine   nine   nine   nine   nine
nine

10 ten   ten   ten   ten   ten   ten
ten

MORE AND LESS

In each row, count the number of items and color the image which has less number of objects.

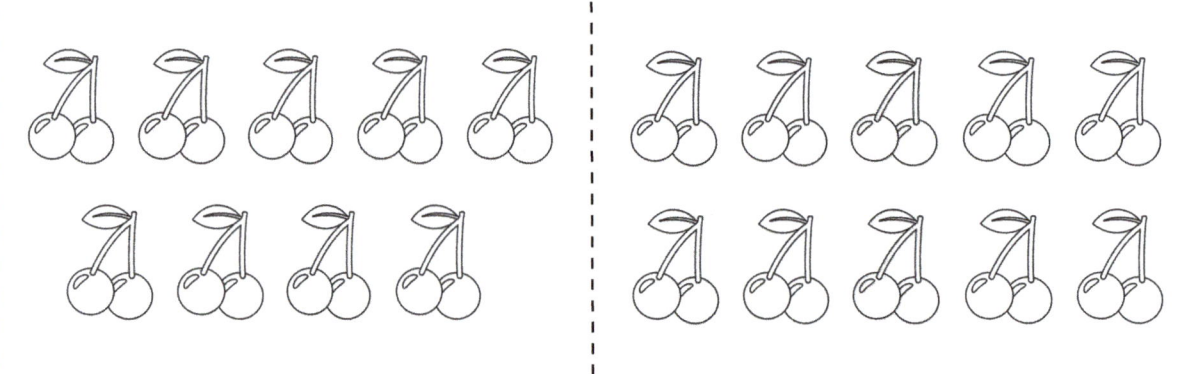

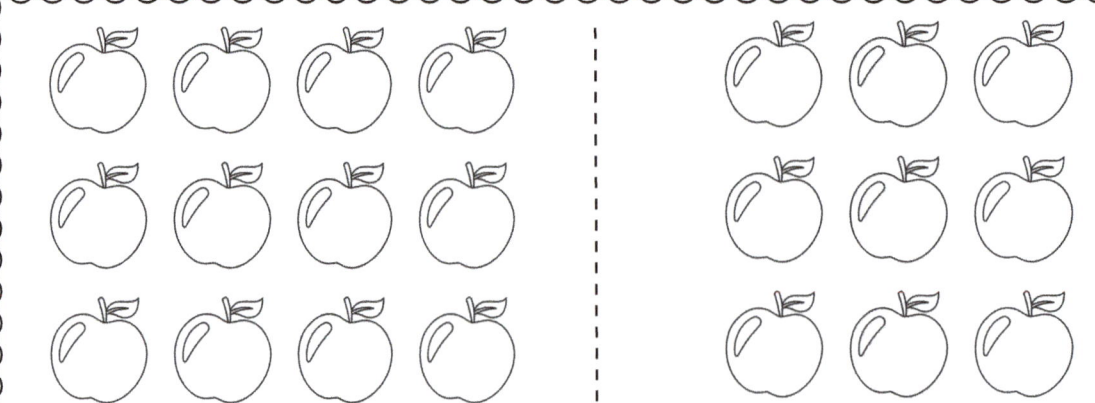

SAME AND DIFFERENT

Circle the items that are the same. Place a square around the items that are different.

COUNT THE ITEMS

Count each items and circle the correct number.

11

8

7

5

12

3

4

1

8

WE ARE WORKING REALLY HARD. GREAT JOB! DON'T GIVE UP NOW...

WAY TO GO!
YOU ARE SUCH A GREAT MATH
DETECTIVE IN TRAINING.
NOW LET'S GET BACK TO THOSE
PROBLEM-SOLVING SKILLS.

SUBTRACT THE PICTURES

Count, subtract, and write the correct number in the empty box.

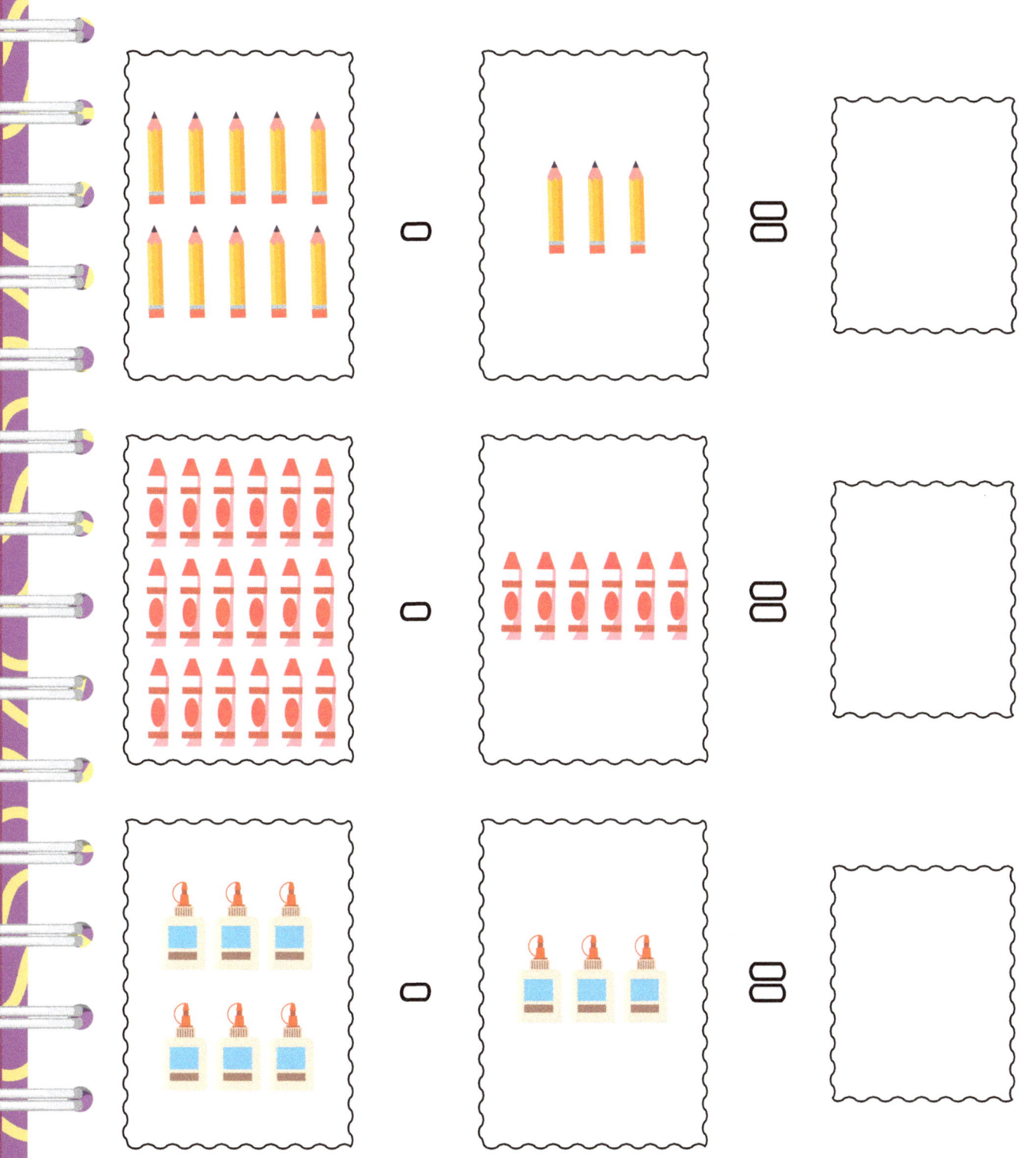

IT'S TIME FOR OUR FIRST CASE!

Try writing or spelling out the numbers we learned in our Great Math Detective warm-up.

1 one

2 two

3 three

4 four

5 five

IT'S TIME FOR OUR FIRST CASE!

Try writing or spelling out the numbers we learned in our Great Math Detective warm-up.

6 six

7 seven

8 eight

9 nine

10 ten

GREAT WORK!

11 eleven eleven eleven eleven

eleven

12 twelve twelve twelve twelve

twelve

13 thirteen thirteen thirteen

thirteen

14 fourteen fourteen fourteen

fourteen

15 fifteen fifteen fifteen fifteen

fifteen

16 sixteen sixteen sixteen sixteen

sixteen

17 seventeen seventeen seventeen

seventeen

18 eighteen eighteen eighteen

eighteen

19 nineteen nineteen nineteen

nineteen

20 twenty twenty twenty twenty

twenty

44

MULTIPLY THE PICTURES

Count, multiply, and write the correct number in the empty box.

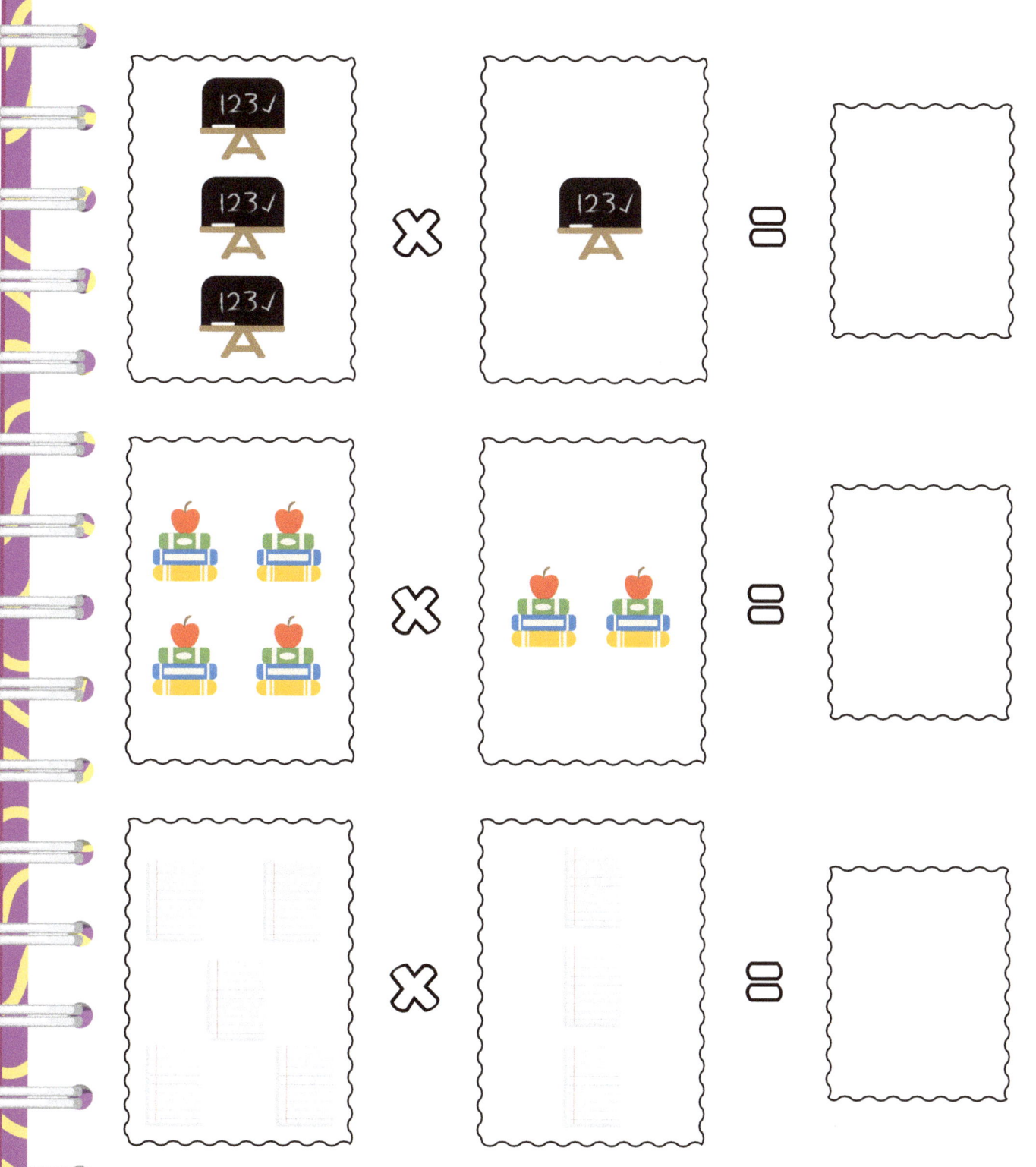

NEW WORD 🚨 ALERT 🚨
NEW WORD 🚨 ALERT 🚨

PERSEVERANCE

The Chief of Math Detectives taught us a new word from *merriam-webster.com.*
The definition of perseverance is

"Continued effort to do or achieve something despite difficulties, failure, or opposition: the action or condition or an instance of persevering"

IT WAS CHALLENGING FOR ME TO DO...

I AM
PROUD THAT
I DID NOT GIVE UP
AND PRACTICED
PERSEVERANCE.

I AM PROUD TO BE A GREAT MATH DETECTIVE BECAUSE...

CERTIFICATE OF COMPLETION
FROM CHIEF MATH DETECTIVE

For :

_____

Awarded the :

Day of : _____

Signed : _____

ANSWER KEY

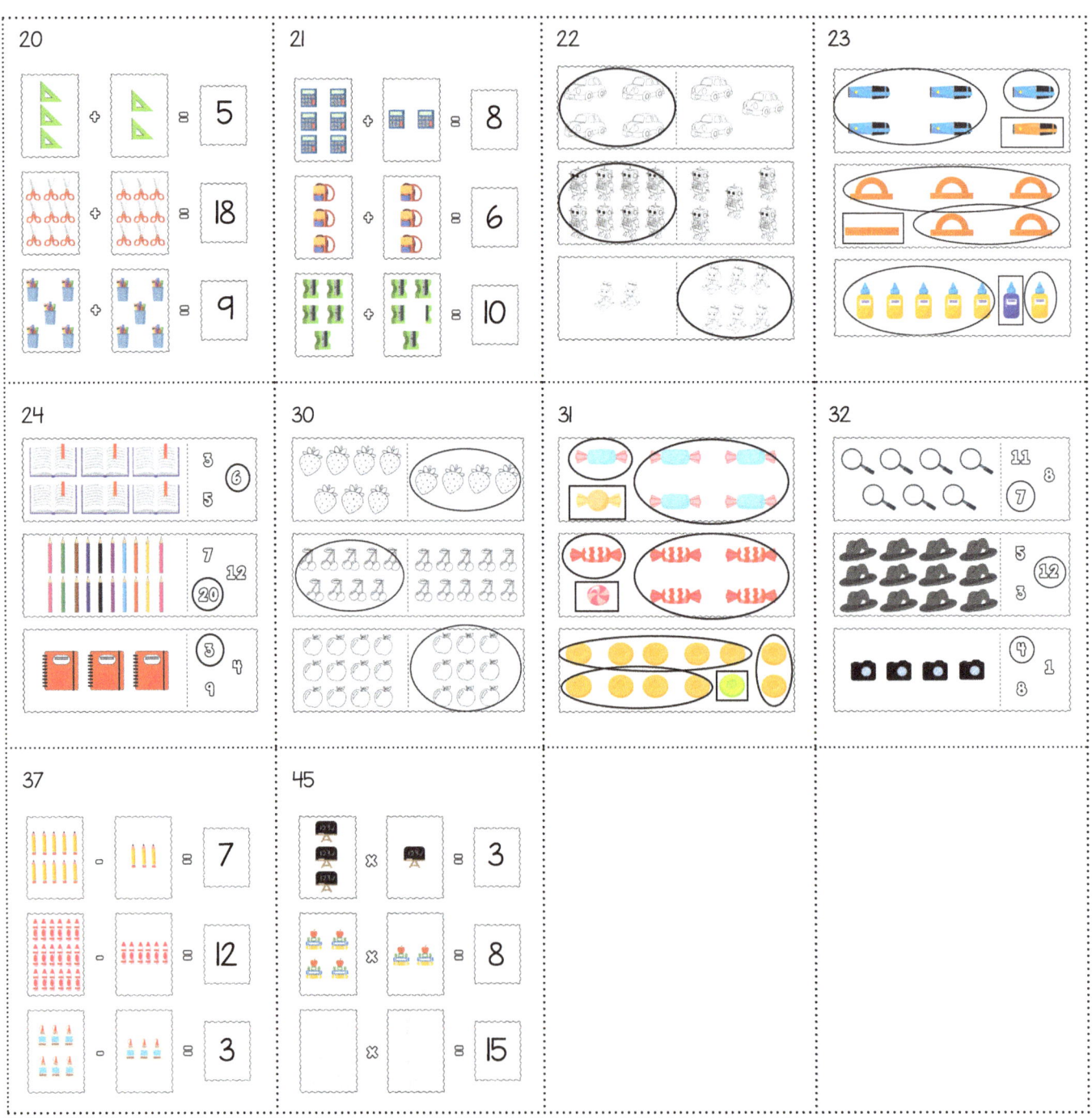

A Special Message from the Chief Math Detective

Congratulations!
Great Math Detective! You've done an amazing job solving puzzles,
cracking codes, and uncovering hidden treasures. Your hard work
and perseverance have made you a true math superstar!

Keep listening, learning, and doing your best in everything you do.
The adventure doesn't end here!

For more exciting challenges and fun, have your grown-up
Like, Follow, and Subscribe:

@maternitymotivation1203

@maternitymotivation_publishing

Join all of our Rockstar Readers by subscribing to
"My Super Mom and Her Sidekick Crew" on YouTube for even more adventures!

Keep shining, and remember, the world needs great math detectives like you!